SCIENCE MILLIE

GROWING LIKE A FLOWER

by Brooklyn Collins

Sunshine & STEM
Read. Explore. Discover.

SCIENCE TIME!

It is Millie's favorite time!
Science time!
Today, Millie is going to learn how a seed grows into a flower with yoga.

First, Millie is a little seed.
This little seed needs water to grow.
Millie curls on the ground to wait patiently
in the dirt for rain.

1... 2... 3...
4... 5... 6...
7... 8... 9...
10!

Waiting is hard.
Millie counts to ten,
taking slow, deep breaths.

Finally rain! Germination can begin.
Millie stretches her arms and legs.
The little seed sends roots into the dirt,
searching for nutrients and water.

Millie stretches, reaching for sunlight.
With a little more time,
and a lot more patience,
the little seed sends a shoot upwards.

Now Millie can be a sprout! Standing tall, the little plant has grown out of the ground.

The little plant needs more food!
Millie stretches out her arms and one leg.
Millie is growing leaves!
With leaves, the plant can use sunlight,
water, and air to make its own food.

Millie is now a young plant.
Millie stands strong on one leg.
The little seed has grown a sturdy stem.

Millie is now a growing plant.
Millie stretches her arms up above her head.
The plant's strong stem can now support
a little flower bud.

Millie is a flower bud.
The little flower bud has grown bigger.
Tucked safely inside are beautiful petals.

Slowly, the petals begin to open.
Millie stretches her arms wide.
The flower is blooming.
This flower is ready to be pollinated.
It has created sweet nectar and
an even sweeter smell to attract
pollinators.

Millie is a butterfly. Millie flaps her wings following the sweet scent of blooming flowers.

As butterflies move from flower to flower, pollen is transferred.

With pollen, flowers create seeds.

These new little seeds will become new little plants, and the cycle continues.

The Life Cycle of a Plant
Plants grow, change, and make new plants!
1. Seed
A new plant begins as a tiny seed.
2. Germination
The seed drinks water. A tiny root comes out first, then a sprout. grows up.
3. Seedling
The sprout grows into a small plant with leaves.
4. Growing Plant
The plant grows bigger. It makes more leaves and a strong stem.
5. Flowering Plant
The plant makes beautiful flowers. Flowers help make new seeds.
6. Fruiting Plant
Flowers turn into fruits. Inside the fruit are seeds.
7. Seed Dispersal
Seeds are spread by wind, water, animals, or falling to the ground.
Parts of a Plant
Flower
Makes seeds.
Leaf
Makes food for the plant.
Stem
Supports the plant and carries water and food.
Roots
Take in water and hold the plant in the soil.

Follow Science Millie
with another Yoga Adventure

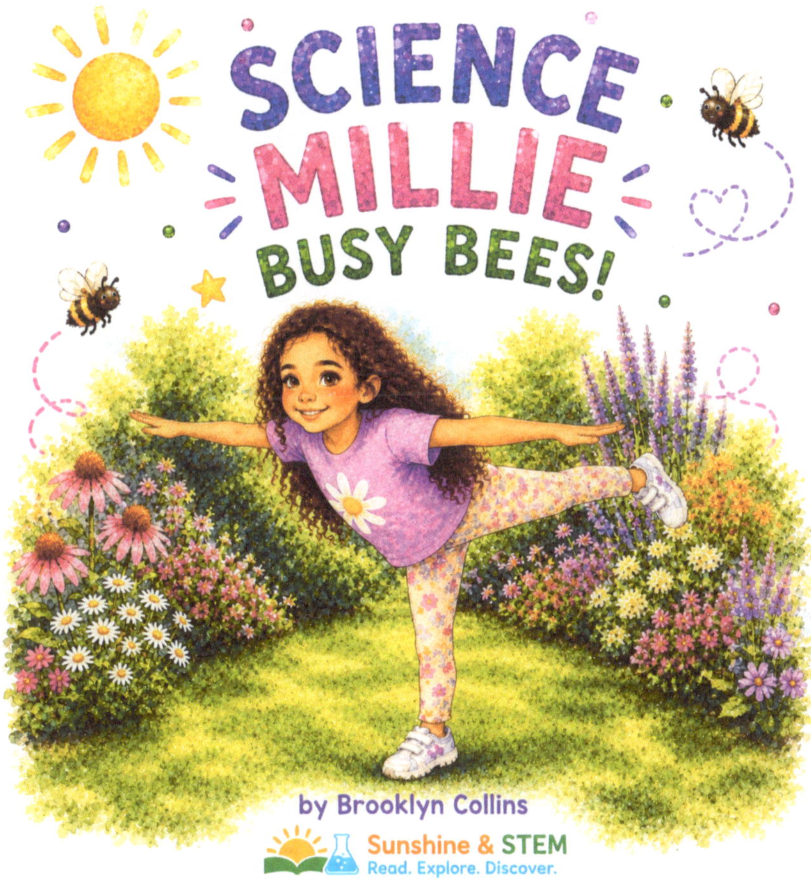

For my little scientists,
whose joy and curiosity
inspire me every day.

Written by Brooklyn Collins
Published by Sunshine & STEM
ISBN: 979-8-951584-00-7

www.ingramcontent.com/pod-product-compliance
Lightning Source LLC
LaVergne TN
LVHW070940210926
848482LV00003B/270
* 9 7 9 8 9 5 1 5 8 4 0 0 7 *